Tornado vs. Hurricane
Author: Jeff Hodson

Copyright © 2025 Jeff Hodson. Published by Roohawk, LLC.
All rights reserved. No part of this publication may be reproduced, distributed, or transmitted in any form or by any means—electronic, mechanical, photocopying, recording, or otherwise—without prior written permission of the publisher, except in the case of brief quotations used for review, educational, or noncommercial purposes as permitted by applicable copyright law.

Publisher's Cataloging-in-Publication Data

Hodson, Jeff.

Tornado vs. hurricane / by Jeff Hodson. — First edition.

A nonfiction science book for young readers comparing tornadoes and hurricanes across 10 categories, exploring which storm is bigger, faster, and more destructive.

Includes bibliographical references.

ISBN 979-8-9988342-3-3 (hardcover)
ISBN 979-8-9988342-4-0 (paperback)
ISBN 979-8-9988342-5-7 (ebook)

1. Tornadoes—Juvenile literature.
2. Hurricanes—Juvenile literature.
3. Natural disasters—Juvenile literature.
4. Severe weather—Juvenile literature.
5. Storms—Juvenile literature.
6. Comparative studies—Juvenile literature.

I. Title.

Library of Congress Control Number: 2025911281

Image Credits

Certain images in this book are provided courtesy of U.S. federal agencies, including the National Aeronautics and Space Administration (NASA), the National Oceanic and Atmospheric Administration (NOAA), and the National Weather Service (NWS). These images are in the public domain and may be used freely.

The inclusion of public domain images does not imply endorsement of this book or its content by NASA, NOAA, NWS, or any other government agency. This book was independently created and is neither sponsored nor approved by any federal agency.

For detailed image credits, please see page 40.

Disclaimer

This book is intended for educational purposes. While every effort has been made to ensure the accuracy of the information contained herein, the publisher and author make no representations or warranties regarding completeness, accuracy, or applicability of the content. Readers are encouraged to consult additional sources when appropriate.

First Edition, 2025

Published by Roohawk Studios, an imprint of Roohawk, LLC
Idaho Falls, ID, USA
roohawk.com

TORNADO VS. HURRICANE

Hey there! I'm Robb, your **Disaster Duel** referee. Let's see which storm comes out on top!

by Jeff Hodson

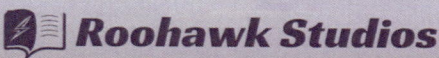

TWO SPINNING GIANTS.
One Epic Battle.

Which is more dangerous, a **tornado** or a **hurricane**?

In this Disaster Duel, tornadoes and hurricanes face off in **10 thrilling rounds**!

LOW DANGER  MAX DANGER!

In each round, the **Danger Meter** shows which disaster is worse. The one with more skulls wins the point!

Will the winner be the storm that causes the **most damage**? Or the one that's **hardest to escape**?

MEET THE DISASTER

Tornadoes

A **tornado** is a small but powerful spinning storm. Some are thin and twisty like ropes. Others are wide, round, and look more like giant funnels.

TORNADO HOT SPOTS

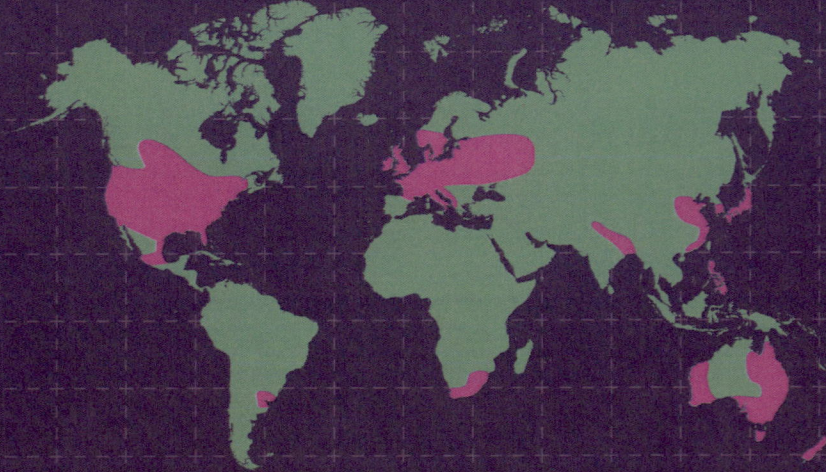

Tornadoes are most common in the United States. There's even a special area called **Tornado Alley**. But tornadoes pop up in many different places all over the world.

Scientists use a special scale to measure how strong a tornado is. It's called the **Enhanced Fujita**, or **EF**, scale.

EF SCALE	
EF-0	Weakest
EF-1	
EF-2	
EF-3	
EF-4	
EF-5	Strongest

The **EF scale** uses the damage a tornado causes to **estimate wind speed**.

Even a small tornado can earn a high rating if its winds cause major damage!

MEET THE DISASTER
Hurricanes

Hurricanes are giant, swirling storms. They form over warm ocean waters near the equator.

Hurricanes often have a calm, clear center called the **eye**.

The **eyewall** circles the eye with the storm's worst wind and rain.

Rainbands of clouds and storms swirl outward from the eyewall.

Like tornadoes, hurricanes are given a number to show their strength. The **Saffir-Simpson Hurricane Wind Scale** ranks hurricanes by wind speed.

CATEGORY
1 — Weakest
2
3
4
5 — Strongest

HURRICANE, CYCLONE, OR TYPHOON?

Hurricanes

Typhoons

Cyclones

Hurricanes, **typhoons**, and **cyclones** are all the same kind of storm. They are all **tropical cyclones**.

People call them different names based on which part of the world they live in.

ROUND 1

Wildest Winds

Will tornadoes or hurricanes score the first point for **fastest winds**?

TORNADO

The air spinning inside a tornado can be super fast! Its winds can reach speeds over **300 miles (480 kilometers) per hour**. That's fast enough to lift cars into the air and snap trees in half!

But most tornadoes have winds below **110 miles (177 kilometers) per hour**.

SPEED SHOWDOWN

Tornado Winds  300 MPH (480 KM/H)

Bullet Train 200 MPH (320 KM/H)

Cheetah 75 MPH (120 KM/H)

HURRICANE

Just like tornadoes, hurricanes can whip up fierce winds. The strongest ones can have gusts over **220 miles (350 kilometers) per hour**.

In 2018, Hurricane Michael's powerful winds flipped this huge ship onto its side.

SCORE

TORNADO	HURRICANE
1	0

Hurricane winds can be fierce. But tornadoes blow them away with the **strongest winds on Earth**! The first point goes to tornadoes.

DISASTER FACT FILE

Animals vs. Storms

TORNADO

Waterspouts are a kind of tornado that forms over water. They can be strong enough to suck up water and maybe even fish!

In some places, people have seen **fish** fall from the sky.

Some scientists think waterspouts might be the reason.

HURRICANE

Hurricanes create **strong waves** that can damage **coral reefs**.

Believe it or not, hurricanes can sometimes help nature in small ways!

Their rain can clean out **lagoons**. They also stir up the ocean, bringing cooler water to the surface. This can cool down the **corals** for a little while.

ROUND 2

Size Showdown

Tornadoes often **look huge in movies**, but are they bigger than hurricanes?

TORNADO

Tornadoes come in all shapes and sizes, but most are actually pretty small.

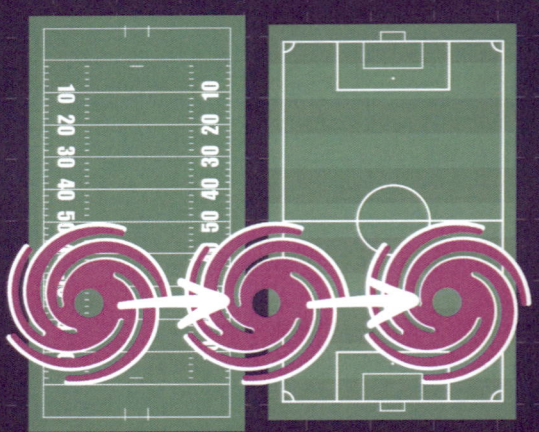

A tornado path is the **trail of damage** it leaves behind. On average, it's only about **50 yards (45 meters) wide**. That's about half the length of a football field!

One of the widest tornado paths stretched **2.6 miles (4.2 kilometers)** across.

HURRICANE

Hurricanes are much bigger than tornadoes. The largest ones can stretch over **1,000 miles (1,600 kilometers) across**.

In 1979, **Typhoon Tip** grew so large that it could stretch from Dallas to New York, or from Barcelona to Oslo.

Most hurricanes are so huge, they can even be seen from outer space.

Hurricanes win this round. And it's not even close!

SCORE

TORNADO	HURRICANE
1	1

ROUND 3

Money Mayhem

Which storm leaves behind the **biggest repair bill**?

TORNADO

Most tornadoes in the United States cause about **$2.5 million** in damage.

In 2011, the Joplin tornado left a jaw-dropping mess. It destroyed over **8,000 buildings** and caused around **$2.8 billion** in damage!

That's enough money to fill up an Olympic-sized swimming pool with dollar bills!

HURRICANE

Tornadoes may sound expensive, but hurricanes usually cost much more. In the United States, an average hurricane causes about **$23 billion** in damage. That's like the cost of 9,200 tornadoes!

In 2005, Hurricane Katrina caused an estimated **$125 billion** in damage. That's like **56 Olympic-sized swimming pools** full of dollar bills!

SCORE

TORNADO	HURRICANE
1	2

Hurricanes win a very expensive point!

ROUND 4

Longest Journey

Hurricanes pull ahead after three rounds! Now it's time to see **which storm travels farther.**

TORNADO

Most tornadoes only travel around **3.5 miles (5.5 kilometers)**. But every once in a while, one goes much farther.

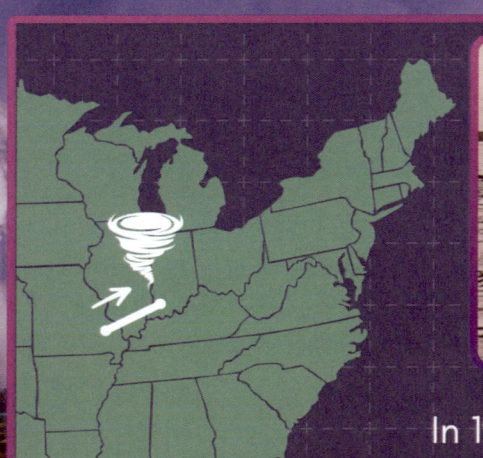

In 1925, the Tri-State Tornado tore across more than **200 miles (320 kilometers)!**

HURRICANE

☠️☠️☠️☠️☠️

It's rare for tornadoes to travel more than **100 miles (160 kilometers)**, but that's a short trip for hurricanes. They can **travel thousands of miles** over water and hundreds of miles across land.

Hurricane Sandy traveled about **1,900 miles (3,000 kilometers)** during its life.

SCORE

TORNADO	HURRICANE
1	3

Hurricanes win this point by a mile, or a thousand!

DISASTER FACT FILE
Wild Weather

A **fire whirl** is like a tiny tornado that is made by fire. In most cases, it's not a true tornado because it doesn't come from a big storm cloud.

But sometimes, a huge, hot wildfire creates its own storm cloud that can make a real **fire tornado**!

A powerful fire tornado from 2018 during the Carr Fire in California.

HURRICANE

In 2017, **Hurricanes Hilary** and **Irwin** got close enough to swirl around each other.

After a few days of circling, both storms faded away over cooler water.

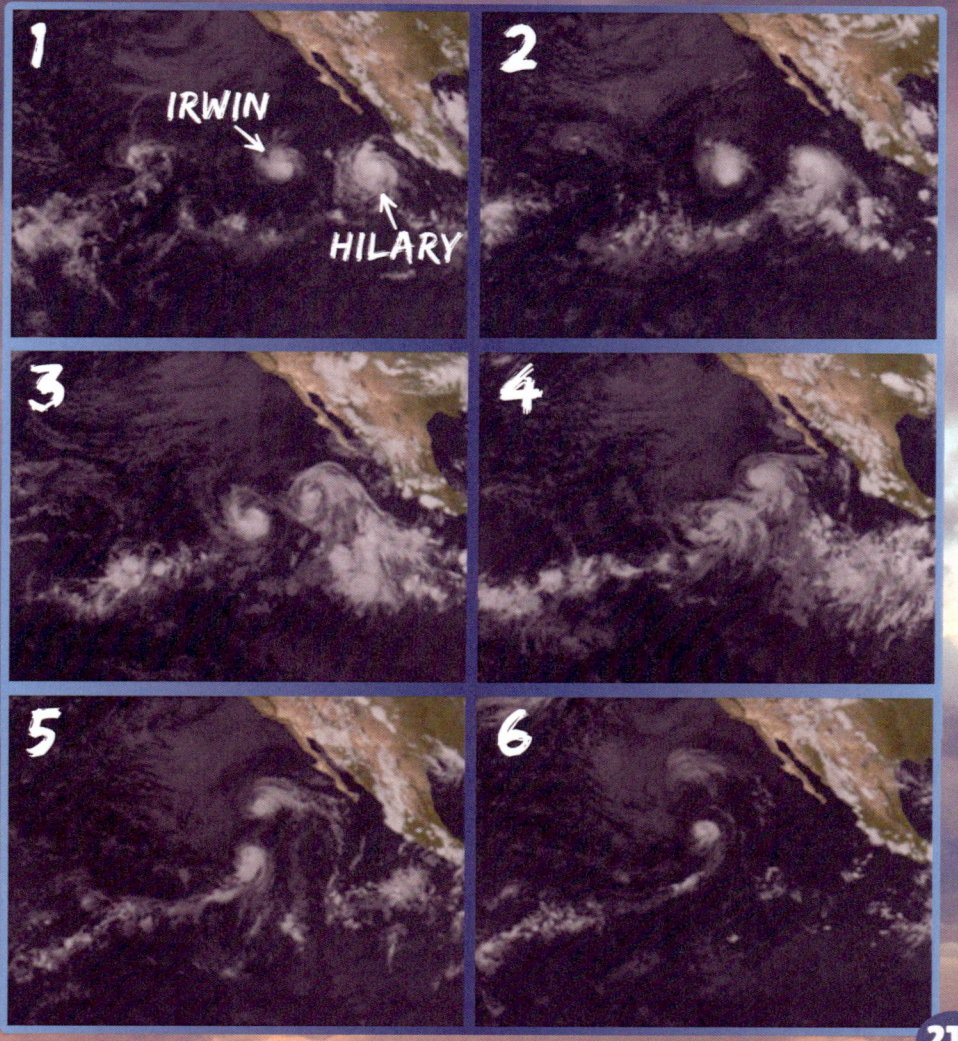

ROUND 5

Race to Escape

Hurricanes have built a strong lead, but can tornadoes catch up? Let's find out **which one is harder to escape!**

TORNADO

Tornadoes might be smaller, but escaping one is not easy. They form suddenly, often with very little warning.

TORNADO SHELTER

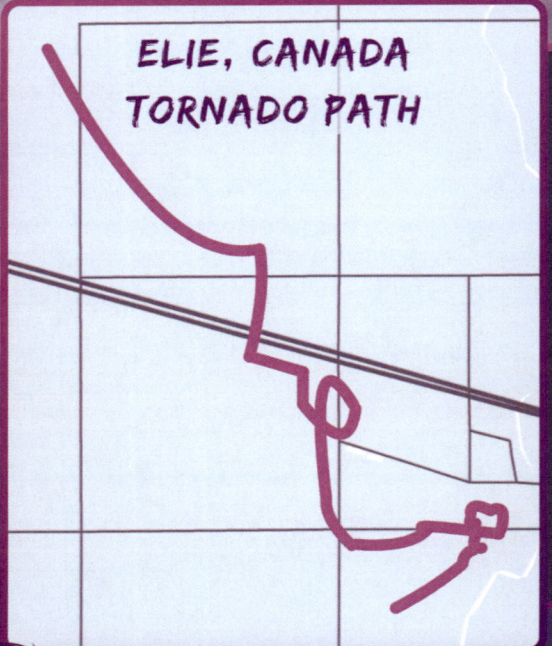

ELIE, CANADA TORNADO PATH

Check out the wild path of the Elie tornado in 2007.

No place near a tornado is truly safe. They can change direction at any moment!

HURRICANE

Hurricanes are huge, but they're also slower. People usually get warnings several days before a hurricane reaches their city.

But long waits, crowded roads, and high costs can make it hard to leave.

☠☠☠☠☠

SCORE

TORNADO	HURRICANE
2	3

If a tornado comes your way, you'll have just a few minutes to find safe shelter. Tornadoes escape this round with a point!

ROUND 6

Last Storm Standing

This windy war is blowing strong! It's time to see **which storm lasts longer**.

TORNADO

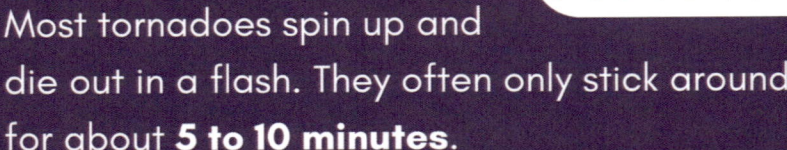

Most tornadoes spin up and die out in a flash. They often only stick around for about **5 to 10 minutes**.

Tornadoes that come from big, powerful storms (called **supercells**) can last for over an hour!

Tornadoes can strike any time, day or night. The most common time is in the evening, **between 4:00 and 9:00 p.m.**

HURRICANE

A hurricane can last for days or even weeks. They keep getting stronger as long as they are over warm ocean water.

In 2023, Cyclone Freddy lasted an incredible **36 days**! It started near Australia and traveled all the way to southern Africa.

Hurricanes hang around long enough to win this point!

SCORE

TORNADO	HURRICANE
2	4

ROUND 7

Storm Race

Tornadoes may have the fastest winds, but which storm would **win in a race**?

TORNADO

Some tornadoes barely move at all, but some can race across the land at **60 to 70 miles (95–113 kilometers) per hour**. That's about the speed of a car on the highway!

Most tornadoes only travel as fast as a horse, about **30 to 50 miles (48–80 kilometers) per hour**.

HURRICANE

Sometimes hurricanes stall over an area for days! But they usually creep along at **10 to 20 miles (16–32 kilometers) per hour**.

ARE YOU FASTER THAN A HURRICANE?

 15 MPH (24 KM/H)

 12 MPH (19 KM/H)

 3 MPH (5 KM/H)

SCORE
TORNADO	HURRICANE
3	4

Some hurricanes can reach speeds above **40 miles (65 kilometers) per hour,** but it's not enough.

Tornadoes finish this race first to grab another point!

ROUND

Deadly Match-Up

Are tornadoes or hurricanes **more dangerous to people**?

TORNADO

Tornadoes can be very dangerous, but they only cause about **80 deaths per year** in the United States.

Europe is another tornado hot spot, but they only have about **10 to 15 tornado deaths** each year.

The Tri-State Tornado of 1925 is known as one of the deadliest. Around **695 people** lost their lives during the storm.

HURRICANE

In the United States, an average hurricane causes about **24 deaths**.

But in other parts of the world, hurricanes can be far more deadly.

In 1970, Cyclone Bhola caused between **300,000 and 500,000** deaths.

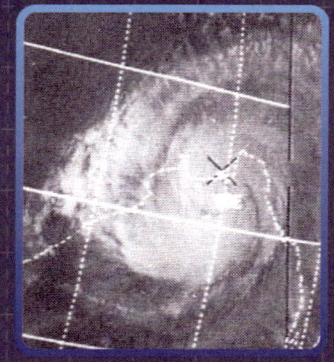

ASIA

SCORE

TORNADO	HURRICANE
3	5

Every storm is a reminder of why safety matters. Hurricanes score big in this deadly round!

DISASTER FACT FILE
Tornado Tag Team

Hurricanes can bring strong winds and flooding.

But did you know they can also make tornadoes?

In 2004, Hurricane Ivan caused more than **100 tornadoes** across Florida in just 3 days!

Hurricanes usually spin up tornadoes near the coast. Most happen from **12 hours before landfall** to **24 hours after**.

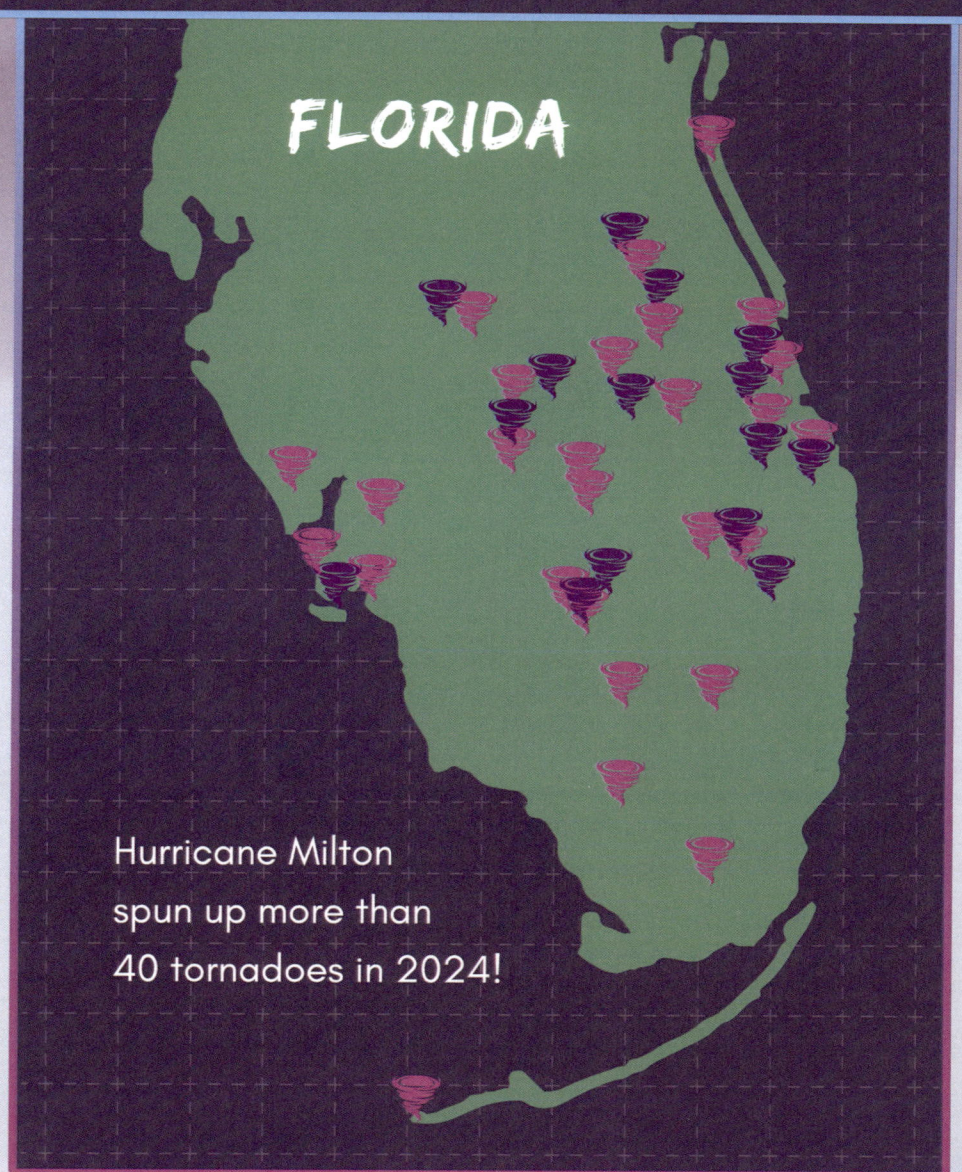

FLORIDA

Hurricane Milton spun up more than 40 tornadoes in 2024!

ROUND 9

Shocking Storms

This Disaster Duel is charging into the final rounds. Let's find out which storm **really brings the boom**!

TORNADO

Tornadoes don't make lightning, but the thunderstorms that create them do!

Supercell thunderstorms produce the strongest tornadoes. They can unleash hundreds of lightning flashes each minute.

Some supercell storms have reached around **500 lightning flashes per minute**!

HURRICANE

Hurricanes don't usually bring much lightning. But when a hurricane hits land or mixes with other weather, it can light up the sky.

Typhoon Matmo was a rare storm. In just one hour, lightning struck 4,400 times. That's about **74 lightning flashes per minute!**

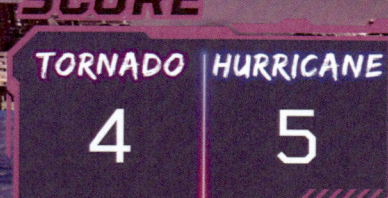

SCORE
TORNADO | HURRICANE
4 | 5

If you want an epic light show, tornado supercell storms are the real winners. Tornadoes score a point for lightning!

ROUND 10

Splash Showdown

The final round is **all about water**. Can tornadoes flood the scoreboard and force a tie?

TORNADO

A tornado's thunderstorm might drop **1 to 3 inches (2 to 8 centimeters)** of rain.

These storms can cause small floods and create hail the size of baseballs!

HURRICANE

Hurricanes often dump **10 to 20 inches (25 to 50 centimeters)** of rain as they move across the land.

Storm surge is often the most dangerous part of a hurricane. It can flood entire coastal towns and sweep away cars and even houses!

Hurricanes take the point in this wet and wild round. The strongest ones can bring several feet (over a meter) of rain. Plus a wall of seawater!

AND THE WINNER IS...

SCORE

TORNADO	HURRICANE
4	6

Tornadoes score big for speed and wind power. But in the end, **hurricanes win this Disaster Duel!**

KEEP THE DUEL GOING!

Ask your parent, teacher, or another grown-up to visit our website!

You'll find:

- More **wild facts** about tornadoes and hurricanes.
- News about **upcoming books** in the Disaster Duels series.
- **Free activities** to use at home or in the classroom.

Parents and teachers can also join the Roohawk Reading Club for the latest **updates** and **free resources**!

Scan the QR code below or go to **roohawk.com/tornado-hurricane**.

Sources

Atlantic Oceanographic and Meteorological Laboratory. (2015, November 13). *45th Anniversary of the Bhola Cyclone*. National Oceanic and Atmospheric Administration. https://www.aoml.noaa.gov/hurricane_blog/45th-anniversary-of-the-bhola-cyclone/

Baird, C. S. (2013, April 30). *Can it rain fish? Science Questions with Surprising Answers*. https://www.wtamu.edu/~cbaird/sq/2013/04/30/can-it-rain-fish/

Beven, J. L. II, Alaka, L., & Fritz, C. (2025, March 31). *Tropical Cyclone Report: Hurricane Milton (AL142024)*. National Hurricane Center. https://www.nhc.noaa.gov/data/tcr/AL142024_Milton.pdf

Chinchar, A. (2023, December 14). *Tornadoes around the world*. Royal Meteorological Society. https://www.rmets.org/metmatters/tornadoes-around-world

Edwards, R. (2013, November). *Tropical cyclone tornadoes: History and state of understanding*. Paper presented at the XIX Jornades de Meteorologia Eduard Fontserè, Cosmocaixa, Barcelona. https://www.spc.noaa.gov/publications/edwards/barcatct.pdf

Heron, S., Morgan, J., Eakin, M., & Skirving, W. (2008). *Hurricanes and their effects on coral reefs*. NOAA's Coral Reef Conservation Program. https://www.coris.noaa.gov/activities/caribbean_rpt/SCRBH2005_03.pdf

Leary, L. A., & Ritchie, E. A. (2009). *Lightning Flash Rates as an Indicator of Tropical Cyclone Genesis in the Eastern North Pacific*. Monthly Weather Review, 137(10), 3456-3470. https://doi.org/10.1175/2009MWR2822.1

NASA Global Precipitation Measurement. (n.d.). *What is the difference between a tornado and a hurricane?* https://gpm.nasa.gov/resources/faq/what-difference-between-tornado-and-hurricane

NASA Global Precipitation Measurement. (n.d.). *What is the difference between a typhoon, cyclone, and hurricane?* https://gpm.nasa.gov/resources/faq/what-difference-between-typhoon-cyclone-and-hurricane

National Hurricane Center. (n.d.). *Saffir-Simpson Hurricane Wind Scale*. https://www.nhc.noaa.gov/aboutsshws.php

National Severe Storms Laboratory. (n.d.). *Severe weather 101: Tornado detection*. National Oceanic and Atmospheric Administration. https://www.nssl.noaa.gov/education/svrwx101/tornadoes/detection/

National Weather Service. (n.d.). *The Enhanced Fujita Scale (EF Scale)*. https://www.weather.gov/oun/efscale

National Weather Service Blacksburg, VA. (2014, September 17). *Hurricane Ivan's tornadoes: 10 years ago*. https://www.weather.gov/media/rnk/past_events/Ivan_Summary_10th_Anniv.pdf

National Weather Service Louisville, KY. (n.d.). *Tornadoes: Frequently asked questions about the power of nature*. https://www.weather.gov/lmk/tornadoesfaq

National Weather Service Milwaukee/Sullivan, WI. (n.d.). *Severe weather awareness - tornado classification and safety.* https://www.weather.gov/mkx/taw-tornado_classification_safety

National Weather Service Paducah, Ky. (n.d.). *Severe weather safety guide: Tornadoes.* https://www.weather.gov/media/pah/Skywarn/SEVERETHUNDERSTORMsafety.pdf

NOAA National Centers for Environmental Information. (2012, January 19). *Monthly tornadoes report for annual 2011.* https://www.ncei.noaa.gov/access/monitoring/monthly-report/tornadoes/201113

NOAA National Hurricane Center & Central Pacific Hurricane Center. (n.d.). *Hurricane preparedness - hazards.* https://www.nhc.noaa.gov/prepare/hazards.php

NOAA Office for Coastal Management. (2025, June 1). *Hurricane costs.* https://coast.noaa.gov/states/fast-facts/hurricane-costs.html

Storm Prediction Center. (2024, December 16). *The online tornado FAQ.* https://www.spc.noaa.gov/faq/tornado/

Storm Prediction Center. (n.d.). *The 25 deadliest U.S. tornadoes.* National Oceanic and Atmospheric Administration. https://www.spc.noaa.gov/faq/tornado/killers.html

UCAR Center for Science Education. (2025). *Hurricanes are changing.* https://scied.ucar.edu/kids/climate-change/hurricanes-are-changing

Williams, E., Boldi, B., Matlin, A., Weber, M., Hodanish, S., Sharp, D., ... & Buechler, D. (1999). *The behavior of total lightning activity in severe Florida thunderstorms.* Atmospheric Research, 51(3-4), 245-265.

World Meteorological Organization. (2020, November 12). *World's deadliest tropical cyclone was 50 years ago.* https://wmo.int/media/news/worlds-deadliest-tropical-cyclone-was-50-years-ago

World Meteorological Organization. (2023, September 23). *Tropical cyclone forecasting.* https://wmo.int/content/tropical-cyclone-forecasting

World Meteorological Organization. (2023, November 24). *Tropical cyclone naming.* https://wmo.int/resources/wmo-fact-sheets/tropical-cyclone-naming

World Meteorological Organization. (2024, July 2). *Tropical Cyclone Freddy is the longest tropical cyclone on record at 36 days.* https://wmo.int/news/media-centre/tropical-cyclone-freddy-longest-tropical-cyclone-record-36-days-wmo

World Meteorological Organization. (2025). *Tropical cyclone.* https://wmo.int/topics/tropical-cyclone

Young, R., & Hsiang, S. (2024). *Mortality caused by tropical cyclones in the United States.* Nature, 635, 121-128. https://doi.org/10.1038/s41586-024-07945-5

Zhang, W., Zhang, Y., Shu, S., Zheng, D., & Xu, L. (2022). *Lightning distribution in tropical cyclones making landfall in China.* Frontiers in Earth Science, 10, 940205.

Zhang, W., Zhang, Y., Zheng, D., & Zhou, X. (2012). *Lightning Distribution and Eyewall Outbreaks in Tropical Cyclones during Landfall.* Monthly Weather Review, 140(11), 3573-3586. https://doi.org/10.1175/MWR-D-11-00347.1

Image Credits

Page 4: Image by Dan Ross, used under license from Depositphotos.
Page 5: Image courtesy of NASA/Goddard Space Flight Center. Public domain.
Page 7: Image courtesy of MassDOT/Jeff Mullan. Public domain.
Page 8: Image by ESA/A.Gerst, retrieved from https://flic.kr/p/MfhNp4, licensed under CC BY-SA 2.0. Some changes applied.
Page 10: Tree Snapped in Half from Tornado. Image courtesy of wrightbrosfan. Public domain.
Page 11: Capsized Ship. Image courtesy of NOAA. Public domain.
Page 12: Image by Rob Atherton, used under license from Depositphotos.
Page 13: Coral Reef. Image courtesy of NOAA. Public domain.
Page 14: Image by Daniel Rodriguez, retrieved from https://flic.kr/p/eAh3rs, licensed under CC BY 2.0. Some changes applied.
Page 16: Joplin Tornado Damage. Image courtesy of U.S. Army Corps of Engineers, Kansas City District and John Daves. Public domain.
Page 17: Hurricane Katrina Damage. Images courtesy of NOAA. Public domain.
Page 18: Image courtesy of the Posey County Historical Society, used with permission. Retrieved from https://www.weather.gov/pah/1925tornado
Page 20: Fire Whirl. Image courtesy of US Fish and Wildlife Service. Public domain.
Page 20: Carr Fire Tornado. Image courtesy of California Department of Forestry and Fire Protection. Public domain.
Page 21: Hurricanes Hilary and Irwin. Images courtesy of National Weather Service (NWS). Public domain.
Page 24: Image courtesy of Mike Coniglio/NOAA NSSL. Public domain.
Page 26: Doppler on Wheels. Photo by Vortex II, retrieved from https://flic.kr/p/8GDS1m, licensed under CC BY 2.0. Some changes applied.
Page 27: Image courtesy of NASA/Randy Bresnik. Public domain.
Page 28: Two Tornadoes. Image courtesy of NOAA Legacy Photo; OAR/ERL/Wave Propagation Laboratory. Public domain.
Page 28: 1925 Tri-State Tornado Destruction. Image by NOAA/NWS Archives. Public Domain.
Page 29: Image courtesy of NOAA – Mariners Weather Log. Public domain.
Page 31: Image courtesy of Illya Tsemenko, used with permission. Retrieved from https://windychi.com/photo/matmo_lg/
Page 32: Baseball Hail. Image courtesy of NOAA Photo Library, NOAA Central Library; OAR/ERL/National Severe Storms Laboratory (NSSL). Public domain.
Page 34: Highway Tornado during Hurricane Ivan. Image by Florida Department of Transportation (FDOT). Public Domain.
Page 34: Truck and Tornado during Hurricane Ivan. Image by U.S. Navy/Jacqui Barker. Public Domain.

www.ingramcontent.com/pod-product-compliance
Lightning Source LLC
Chambersburg PA
CBRC100102100526
44582CB00011B/169